Dark Matter, Unified Field Theory, and UFO's, Are Understandable and Achievable.

Dark Matter, Unified Field Theory, and UFO's, Are Understandable and Achievable.

Bob Šablatúra

2nd Edition

Library of Congress Control Number: 2015900413
ISBN: Hardcover   978-1-5035-0140-9
                  Softcover   978-1-5035-0141-6
                  eBook   978-1-5035-0142-3

Print information available on the last page.

Rev. date: 03/27/2021

To order additional copies of this book, contact:
Xlibris
AU TFN: 1 800 844 927 (Toll Free inside Australia)
AU Local: 0283 108 187 (+61 2 8310 8187 from outside Australia)
www.Xlibris.com.au
Orders@Xlibris.com.au
698263

CONTENTS

I wish to dedicate this book to my kids,
Eryn and Joshua.

CHAPTER 1

Gravity and Photons

We find from Albert Einstein's and Hendrik Lorentz's equations that the mass of an object is altered as it approaches the speed of light. [1]

In fact, the closer the velocity of the object comes to c, the more massive it becomes. This is called mass dilation and is represented mathematically as:

$$m = \frac{m_0}{\sqrt{1 - \dfrac{v^2}{c^2}}},$$

where m = relativistic mass of particle, m_0 = rest mass of particle, v = velocity of the particle relative to a stationary observer, and c = speed of light.

Graph 1: Mass increase

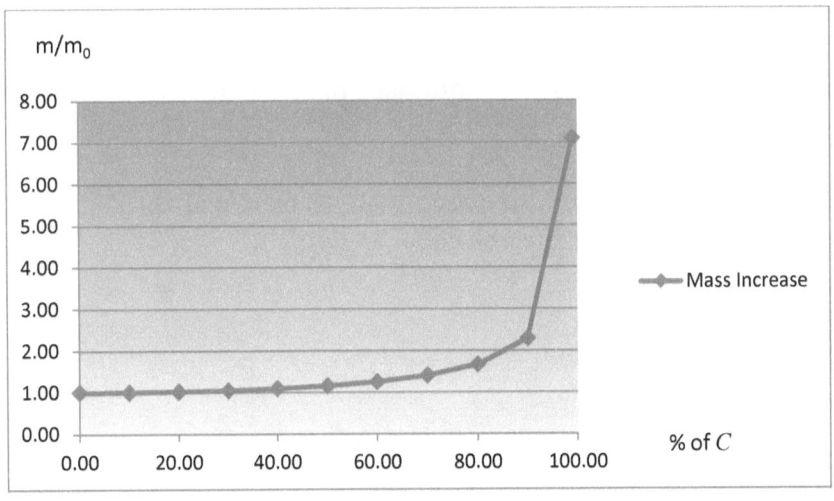

If we look at this as a graph, we can see that at speeds close to the speed of light the mass increases rapidly, and as it approaches closer and closer towards c (but never getting to c) its mass continues to increase.

The end result is that as the particle approaches the speed of light its mass becomes infinite. (see Table 1)

Table 1: Percentage speed of light versus the increase of mass:

% of C	Mass increase m/m_o
10.00	1.01
50.00	1.15
90.00	2.29
99.0	7.09
99.90	22.37
99.99	70.71
99.999	223.61
99.9999	707.11
99.99999	2236.07
99.999999	7071.07
99.9…9	∞

Therefore, it is impossible for ordinary matter to reach the speed of light (c).

But a photon travels at the speed of light or c, and for that to be possible, its rest mass must be equal to zero.

Photon rest mass (m_p) = 0
Relating to mass dilation equation:
∴ $m_o = m_p = 0$.

Therefore, if we now wish to calculate the net weight force on a photon or the effect gravity will have on a photon, we find the following:

Weight (w) = mass × gravity (g), where g = individual gravity of a body.

$$\text{But } m_p = 0$$

$$\therefore \quad W = 0 \times g$$

$$W = 0$$

Thus, the influence of any star on the motion of a photon is zero.

Therefore, it is impossible for any star's gravity to alter the passage of a photon. There must be another cause for the change in path of a photon as it passes a star.

The theory I wish to propose is related to the mechanism by which glass or other media can alter the speed and direction of a photon. This I will endeavour to explain elsewhere.

Note: There are experiments that use lasers directly to try and detect and measure gravitational waves. These devices are bogus, simply because photons themselves are unaffected by the influence of gravity.

DARK MATTER, UNIFIED FIELD THEORY, AND UFO'S, ARE UNDERSTANDABLE AND ACHIEVABLE.

11

CHAPTER 2

Dark Matter Equates to Photons and Negative Mass see (Ch 8.)

Craig F. Bohren showed 'how a particle can absorb more than the light incident on it' (*American Journal of Physics* 51(4) April 1983, p.325). In his experiment, aluminium was exposed to electromagnetic radiation of 8.8 electron voltages, subsequently the atoms of aluminium and their respective electrons synchronised oscillations with the incoming radiation and resulted in the incoming radiation being absorbed by Aluminium Atoms. [2] The net effect allowed the Aluminium atoms to absorb photons from great distances from the parent metal substrate, compared to other electron voltages and frequencies of incoming radiation. (Bohren, 1983)

In recent years, man has been teleporting photons over ever-increasing distances. This work has been supported by Shuntaro Takeda et al, in their work "Deterministic quantum teleportation of photonic quantum bits by a hybrid technique", (*Nature,* 500, 315–318, 15 August 2013, doi: 10.1038/nature12366, published online 14 August 2013).[3]

The Red Shift of Hydrogen Spectra of the electromagnetic radiation shows that as a photon moves through the vacuum of space, and gets older, it loses energy, shifting slowly from high to low energy (i.e. blue light changes into red light). [4] (Hydrogen shift)

Based on this information, a possible explanation for some of the dark matter in the universe is that it is comprised of photons, and they are spread throughout the universe. Since photons do not possess any rest mass (rest mass = 0), it is extremely difficult to locate them. Therefore, as the photons lose energy they can be teleported, absorbed, and bound to other particles until they can be radiated once more. Photons which lack the energy to exist freely, they will be absorbed by matter elsewhere. In other words, quantum absorption can occur; this is where a photon's location is not essential, and the photons can be teleported to the nearest object, irrespective of the distance.

A simple experiment, as an example of this is to pass an electrical current through a filament in a light bulb. As the current increases in the wire, it excites the atomic structure and that in turn causes oscillations of electric charges within and between atoms. The net effect is to create synchronised and unsynchronised resonance on a quantum level, which can either attract or repel photons.

Note: Dark matter could be comprised of negative mass, photons, neutrinos, planks sized particle dust, spread out throughout the universe.

Physicist Dr Mark Raizen and his team at the Centre for Nonlinear Dynamics and Department of Physics at the University of Texas at Austin focused a laser beam on to a glass bead to suspended it in the air. The net effect was to make it appear as if it was levitating. This experiment showed that the momentum of photons was real and that it can interact with matter.

The reason why the clear glass bead was levitated is because of electrostatic charges associated with the photon interacting with the charges that exist with atoms and molecules alike. These interactions create a dragging effect on the photons and in turn a loss of energy which imparted to the surrounding matrix of electrons and protons. The net effect of this interaction causes the glass bead to one heat up and two to push up on the bead and cause it to levitate.

This interaction of charges from photons with the ionic wind of stars is what causes lensing or bending of electromagnetic radiation, as it passes stars. It is also the cause of refraction of photons from one material to another and it also the reason why photons slow down when they move in progressively denser materials. The reverse happens seeing photons speed up when they move from a very dense to a lesser dense medium.

DARK MATTER, UNIFIED FIELD THEORY, AND UFO'S,
ARE UNDERSTANDABLE AND ACHIEVABLE.

13

CHAPTER 3

Dimensions

The following theory of 'dimensions' is a little difficult to explain, but I will try my best. Among many physicists today, there is the belief that there exists eleven 'dimensions' and that each one of these dimensions builds upon the previous ones. Each one of these dimensions is at right angles to all the previous ones, [5] and it must be noted that these dimensions are active and are present at all times. In other words, the second 'dimension' cannot exist without the existence of the first. There is a strong interdependence between each one of the dimensions.

You cannot have the existence of a higher order dimension without the previous ones. A good analogy for this would be the number system. You have the numbers 1, 2, 3, 4, 5, 6, 7, 8, 9 . . . etc. But you cannot have the number 2 unless you have the number 1; nor can you have the number 3 unless you have the number 2. This can be applied to every number in the counting system. Therefore, the existence of any one number (or 'dimension') depends solely on the existence of its forerunner, for without it you cannot have the next one.

Since our human bodies and our consciousness exist in the first three dimensions, we are all genetically and biologically programmed by our environment to see only the first three dimensions. That is why we can only see that which is around us. We cannot see the other dimensions. Yet if we look carefully, we can observe the effects of these other 'dimensions' on our everyday lives.

To continue, I must make another analogy and that is each 'dimension' can be related to a plane of existence or the cross product of other planes of existence. Each plane of existence/ 'dimension' can be related to a vector.

The First Dimension

The first dimension is a straight line that goes on forever. It has no thickness or depth to it, for as soon as you add anything to it, it ceases to be one-dimensional. See figure 1.

Figure 1: A line that goes on forever

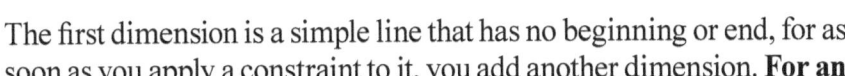

The first dimension is a simple line that has no beginning or end, for as soon as you apply a constraint to it, you add another dimension. **For an action to exist or take place, without any other construct, is a Force.**

The Second Dimension

The Second dimension is a line at right angles to the First Dimension and which cuts the first. The act of cutting a one-dimensional line, you have introduced an action, which is at right angles to the first, therefore creating something new.

With the actual act of cutting the first dimension, we have applied constraints to it, giving it a length and breadth or width. We have now a two-dimensional object.

The second dimension now has two components: a length (L) and a breadth (B). A simple example of (L) × (B) is below. See Figure 2.
Figure 2: Two planes of existence or vectors acting at right angles

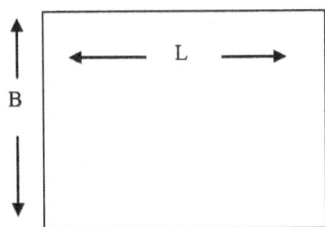

Note that each new dimension is always at right angles or 90° to the previous dimension.

Dark Matter, Unified Field Theory, and UFO's, Are Understandable and Achievable.

15

A partially correct example of this would be an image on a television screen; note that the phosphorus crystals would actually give a third dimension. But if we look at the screen perpendicularly it would appear to be two-dimensional.

The Third Dimension

The third dimension, which is at right angles to the previous two, is created when we impose another action to the second dimension. Since each of these actions has a particular direction and size/ magnitude, they can be referred to as vectors.

A simplistic example of Length (L) × Breadth (B) × Depth (D) is given below.

See figure 3.

Figure 3: Three planes of existence or vectors acting at right angles:

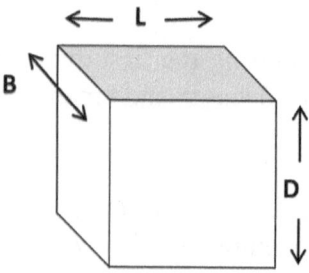

To summarise, the first dimension is a line with no other parameters to it. But in order to change the first dimension, you must then include a second dimension.

Before you can create the third dimension, you must have the previous two dimensions first. **An extremely important point to note is that we cannot have a higher order dimension occurring unless the lower order dimensions are already represented.**

It does not matter where you look, in today's world you will find examples everywhere.

The Fourth Dimension and Gravity

Trying to represent this on paper is tricky, in that you need to create a new plane of existence at right angles to all the previous ones. To simplify everything, I will represent the first three planes of existence (dimensions), or vectors, as a dot on a page, and from that we extrapolate outwards. For a particle with a definite mass (*m*), which will influence the fourth-dimension space in a neutral way, see Figure 4. Gravity is a field effect that exists in the fourth dimension.

Figure 4: Gravity acting at right angles to the first three dimensions

An object with definite mass (*m*)

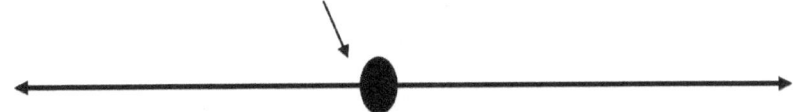

Gravity is a force which attracts all matter towards itself. It works at right angles to all objects which have mass. Equally, the gravitational field of a mass is directly proportional to its mass and it acts at 90 degrees to it.

Gravity (*g*) is directly proportional to the amount of mass in an object.

Mass is the sum of all the protons, neutrons, electrons and other subatomic particles (m_{sub}) in a particular area. Therefore, Mass (*m*) = the sum of ($\sum$) and the modulus of all subatomic particles (m_{sub}) multiplied by their individual specific modulus gravitational values (g_{sub}).

Therefore $$g = \sum |m_{sub}| \times |g_{sub}| \times \sin \theta.$$

Dark Matter, Unified Field Theory, and UFO's,
Are Understandable and Achievable.

17

The intensity of Gravity can also be affected by density, as can be observed in black holes.

$$\text{Density} = \sum | \, m_{sub} \, | / V.$$

If a mass has no net disturbance of the fourth dimension, you would observe nothing happening (see figure 4). Yet if there is an active influence in the fourth plane of existence (4th dimension), the intensity of this gravity (g) is also directly proportional to the distance (d) from the mass.

Gravity (G): $g \; \alpha \; 1/d^2$ from the mass [12]   (α = is proportional to).

The influence of this disturbance will be greatest in four-dimensional space (4D space). Since this disturbance exists outside the first three dimensions, we would not be able to directly perceive it. The only way we can detect it is to observe its influence on objects which are also active in 4D space. In a more simplistic 2D planar view, we could represent it as a potential well; see figure 5.

Figure 5: A gravitational potential well

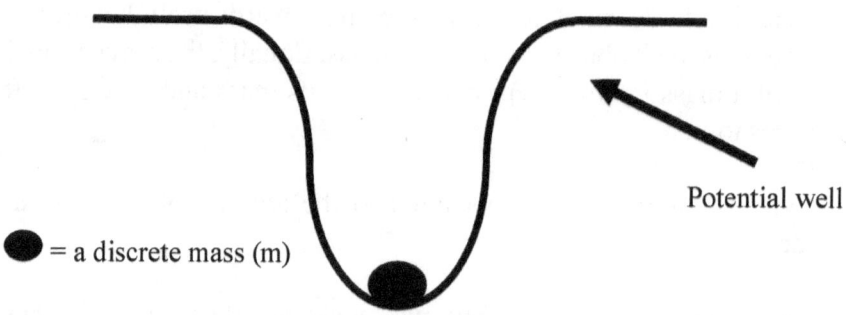

Potential well

= a discrete mass (m)

The $1/d^2$ well simulates the potential well of gravity around the surroundings of a point mass.

The point masses are three dimensional while the potential wells are distortions in the fourth dimension, creating the potential wells as below.

See figure 6.

Figure 6: Relating size of disturbance/mass potential well

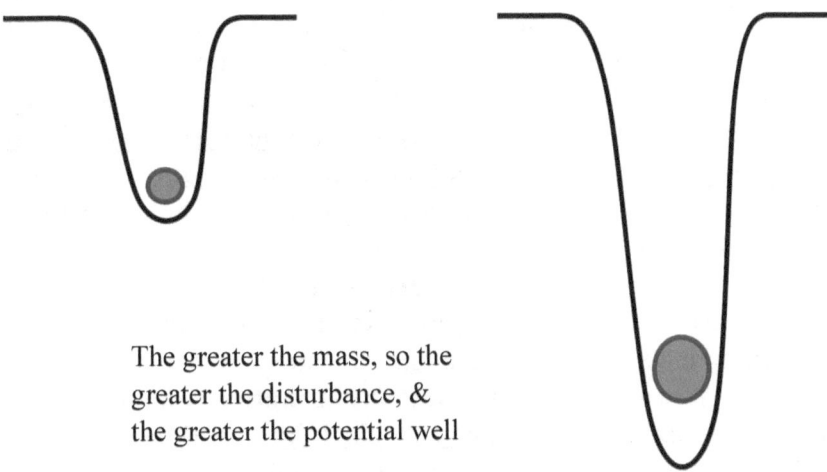

The greater the mass, so the greater the disturbance, & the greater the potential well

A better image could be created if you rotate it about its vertical axis

Black Holes

A black hole's gravitational intensity (g) is directly proportional to the amount of mass in a particular area (density) and the magnitude of this quantity of mass being in one location. It is a sum of all its mass; this includes the addition of all protons, neutrons, electrons, and other subatomic particles (m_{sub}) in a particular area. Therefore, its gravity (g) = the sum of ($\sum$) mass (m) of all subatomic particles (m_{sub}) multiplied by their individual specific modulus gravitational values (g_{sub})/Volume (V in cm^3):

$$\text{Total Mass} = \sum | m_{sub} |/V$$

DARK MATTER, UNIFIED FIELD THEORY, AND UFO'S,
ARE UNDERSTANDABLE AND ACHIEVABLE.

19

Therefore $\qquad g = \{\sum |m_{sub}| / V\} \times |g_{sub}| \times \sin \theta.$

The modulus of mass was taken as tachyons which most likely will have negative mass and negative mass, leading to gravity being repulsive. This will be covered later in this dissertation.

In and around a Black Holes, the Mass and the space between particles is compressed, by gravity, to the point where all particles coexist, or is the same size as plank particles (about a 10^{-38} m^3). Subsequently this has the impact of directly affecting the other dimensions and altering the way they function. (Note, it be should be possible to observe intense electric and magnetic field affects around the equator of a Black hole, conversely at the poles these are weakened). The mass and gravity of a black hole continues to increase, due to it absorbing other bodies in universe. This process of escalating mass and gravity in black holes continues with an attractive value for gravity until the black hole can convert matter into negative mass. Thus, as the black hole becomes more massive, a set amount of matter is being turned in negative mass. (Note, when two black holes collide, a portion of their mass would disappear as it is turned into negative mass.) When a sufficient amount of negative mass builds up within the black hole, the resulting affect is to see the black hole instantly expand with light and faster than Light particles, and in addition to this Plank particles. The result is to send very high energized particles instantaneously everywhere throughout the universe. Note any matter of negative mass will age very quickly and turn into normal matter because of Lorentz's and Einstein's relativity equations. (see chapter 8).

Thus, if you would observe the structure of the universe you should see regions where there is complete blackness and these illusionary balls of darkness should be numerous galaxies. (Note if you should give the impression of Swizz cheese.) Secondary Proof theory could be the existence of an old star that is older than NASA's Wilkinson Microwave Anisotropy Probe (WMAP) estimated age of the universe of 13.772 billion years.

A Galaxy maybe in the order of 1×10^{20} metres in diameter and average Hydrogen may be 1×10^{-10} meters in diameter and a Plank particle is the order of 1×10^{-38} meters; - If we subject a Galaxy to a Black Hole it would be crushed down to the size of 1×10^{-18} meters. This being smaller than an atom and even smaller that a proton of 1.75×10^{-15} meters and an electron of 1.429×10^{-16} meters. See appendix for the calculation of size of electron. If apply this theory to our universe of 13.772 billion years, we could reduce it to a Plank Black Holes $260.55594 \times 10^{-14}$ meters which larger than the size of a proton but smaller than hydrogen atom.

PS If two, Neutron stars should collide, there mass should be conserved and some electromagnetic radiation with subatomic particles being released. But if their combined masses can create a black hole, the end outcome would be completely different. For when two black holes collide a portion of their mass is converted to negative mass. (note: - See in the appendix point 5 for equations relating gravity to electrostatics and appendix point 9 for equations related to the theory about negative mass.)

DARK MATTER, UNIFIED FIELD THEORY, AND UFO's,
ARE UNDERSTANDABLE AND ACHIEVABLE.

21

CHAPTER 4

The Fifth Dimension (Electric Field)

Every plane of existence (dimension) has an effect on every other. If a mass has no net disturbance of the fifth dimension, you will observe nothing happening. Yet there are three ways the dimension can be influenced.

1) The first is neutral and nothing will be observed (see Figure 4).
2) The second is where an object can influence the fifth plane of existence (dimension) in a positive way.
3) The third is where an object can influence the fifth plane of existence (dimension) in a negative way.

The key to the fifth dimension, and electric fields, is the plank particle. Everything is made up of **plank particles**. I hope to **expand on this later** and I will bring it in the study of positron and electron collisions.

All the arguments that I have used in relation to the fourth dimension can be also applied to the fifth dimension.

Electric fields function in the fifth dimension. When you investigate a point charge, it does influence the space around it and the field strength can be calculated by the equation $E = k\, q_1/d^2$, where $E =$ electric field, $q_1 =$ charge, $d =$ distance, and k a constant [7, 8].

You cannot see an electric field directly, but you can observe its effect on other charges and the space around it. When you plot electric field strength (intensity) against distance you can develop a potential well as in Figure 7. This potential well is similar to that of the gravitational one.

Figure 7: Electric charge potential

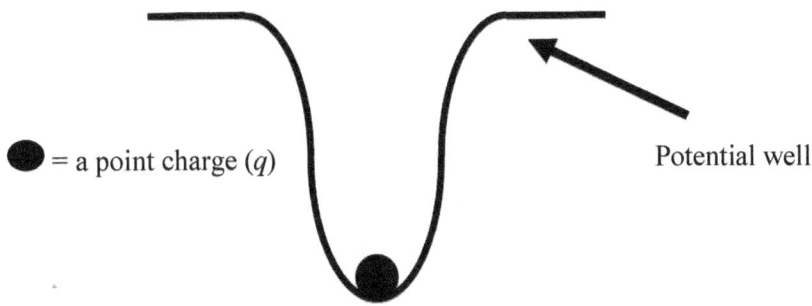

● = a point charge (q)   Potential well

The greatest electric charge disturbance occurs at the mass, while as you increase your distance from the point charge the electric field strength drops with $1/d^2$, just as the disturbance in the fourth dimension.

An electric field cannot exist without either a positive or a negative charge being present. The strength of this electric field is dependent on the magnitude of a mass nearby producing it. [7, 8] The mass is three-dimensional. All **negative charges** affect the fifth dimensional space in one way while conversely **positive charges** affect the fifth dimensional space in a different but related way. The two fields are related to each other but alter the fifth dimension inversely and mirror like to each other.

Like Charges

If you try to put together two like charges, they will repel each other. This is because both these charges are trying to distort the fifth dimension in exactly the same way. As shown in Figure 8, as you try and bring together the two like charges, q_1 and q_2, you find it increasing difficult to do so. The closer they come to each other, the more the force pushes them to move away from one another. The force increases by $F = k\,q_1\,q_2 / r^2$, [7, 8, 9] where F = force, q_1 = charge, r = distance between charges, and k a constant.

DARK MATTER, UNIFIED FIELD THEORY, AND UFO'S,
ARE UNDERSTANDABLE AND ACHIEVABLE.

23

Both like charges want to change the fifth dimension in the exact same way, and this is not possible. Like an over-wound spring, it will tolerate so much before it pushes back. As the initial force is applied inwards, a resultant force is generated, which acts in the opposite direction, pushing both charges q_1 and q_2 outwards and away from each other.

a) The initial force is applied inwards.

b) The initial action is met with an ever-increasing opposing force pushing both charges apart.

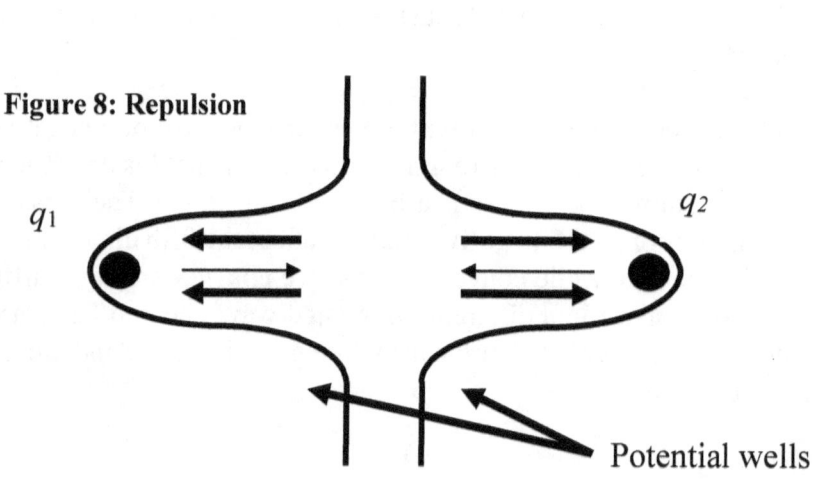

Figure 8: Repulsion

q_1

q_2

Potential wells

● = like point charges (q_1, q_2)

Unlike Charges

If you bring together a negative and positive charge, they attract each other. This is because both these charges are trying to distort the fifth dimension in such a way that they want to cancel each other out.

For as you try and bring together the unlike charges, q_1 and q_2, you find it increasingly easy to do so. The closer they come, the more they want to move towards each other, as the force between them increases by $F = k\,q_1\,q_2/d^2$.

Figure 9: Attraction

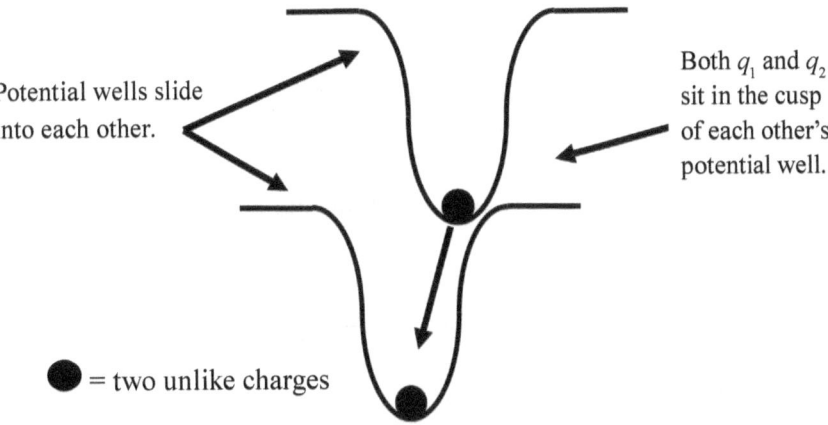

Potential wells slide into each other.

Both q_1 and q_2 sit in the cusp of each other's potential well.

● = two unlike charges

The charges want to achieve a neutral state. See figure 9. They will both slide into each other's potential well. But in doing this, they will partly neutralise each other. A very good example of this is the helium atom, for it appears to not have any charge and is very stable. This can be very easily demonstrated if you bring a helium atom up to a charged metal plate; you see that nothing happens. But if you were to remove one electron from the atom it becomes charged. Thus, if you now bring it up to a charged plate, it will either be attracted or repelled from the charged metal plate, depending on the charge of the plate and the helium atom.

DARK MATTER, UNIFIED FIELD THEORY, AND UFO'S,
ARE UNDERSTANDABLE AND ACHIEVABLE.

25

CHAPTER 5

The Sixth Dimension (Magnetic Field)

Unlike gravity and electric fields, the magnetic component only comes into existence when an electric charge is on the move. [10] For if there is no movement, there is no creation of a magnetic component (Note; - As stated previously all dimensions need to active beforehand). Again, this field is at right angles to the direction of motion of a negative or a positive charge. See figure 10. This property of charges is used with great benefit in the industry to recoup energy from charged particles in chimney stacks, when they pass hot ionised gases through super conductive coils to create electrical energy, to electric motors and generators.

Figure 10: Magnetic Field comes into existence only when a charge moves

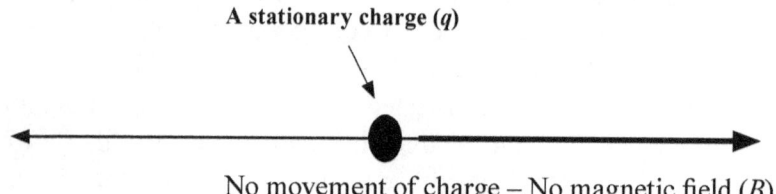

A stationary charge (q)

No movement of charge – No magnetic field (B)

Magnetic fields function in the sixth dimension. The strength of this magnetic field is directly related to the speed and magnitude of the charge that is moving.

There is no such thing as a stable magnetic monopole, as it requires energy to hold it form. The magnetic field is actually a disturbance of the sixth dimension. It is the interaction of **a five-dimensional moving charged object with the sixth dimension that creates the magnetic field.** (See the appendix.)

The magnetic field also creates a potential well. This potential well is similar to that of the gravitational and electric field.

Figure 7:

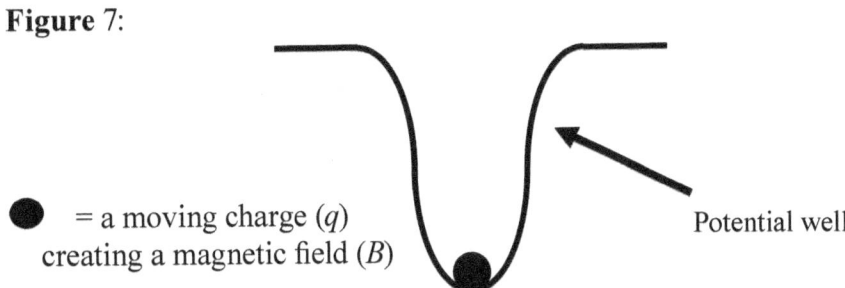

= a moving charge (*q*) creating a magnetic field (*B*)

Potential well

The amazing thing is that these fields have a North and South Pole component. These magnetic poles can be created via either a negative or a positive charge, depending on how the charge is moving. [10] (See the appendix.)

The poles of a bar magnet behave in many ways like charged objects.

When you investigate a magnetic field, it influences the space around it and its field strength (B) decreases directly proportional to the distance (d) from it.

The magnetic field (*B*) α $1/d^2$ (α = 'is proportional to' and *d* = distance from magnetic source).

Now the fascinating part is that the behaviour of like and unlike charges is replicated in like and unlike poles.

Like Poles (North to North or South to South)

If you try putting together two poles of the same polarity, they will repel each other. This is because both these poles are trying to distort the sixth dimension in exactly the same way. (See figure 11.) As you try to bring together the two like poles (pole 1 and pole 2), you find

Dark Matter, Unified Field Theory, and UFO's,
Are Understandable and Achievable.

27

it increasingly difficult to do so. The closer they come to each other, the more the force pushes them away from one another.

Both poles want to change the sixth dimension in exactly the same way, and this is not possible. Like pushing a spring, it will be able to tolerate so much before it pushes back. As the initial force is applied inwards, a resultant force is generated, which acts in the opposite direction, pushing both poles outwards and away from each other.

a) The initial inward applied force is represented by:

⟶

b) The initial action is met with an ever-increasing opposing force, pushing both poles apart, represented by:

⟵

● = a magnetic source

Figure 11: Repulsion

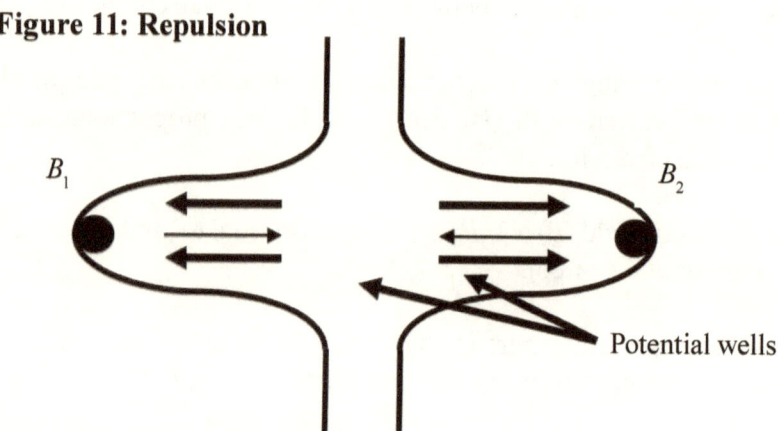

Potential wells

● = like magnetic poles $B_1 = B_2$

Unlike Poles (North to South or South to North)

If you bring together two unlike poles, they attract each other. This is because both these poles are trying to distort the sixth dimension in such a way that they want to cancel each other out. As you try to bring together the unlike poles, you find it increasingly easy to do so. The closer they come, the more they want to move towards each other, as there is a strong force of attraction between them. See figure 12

Figure 12: Attraction

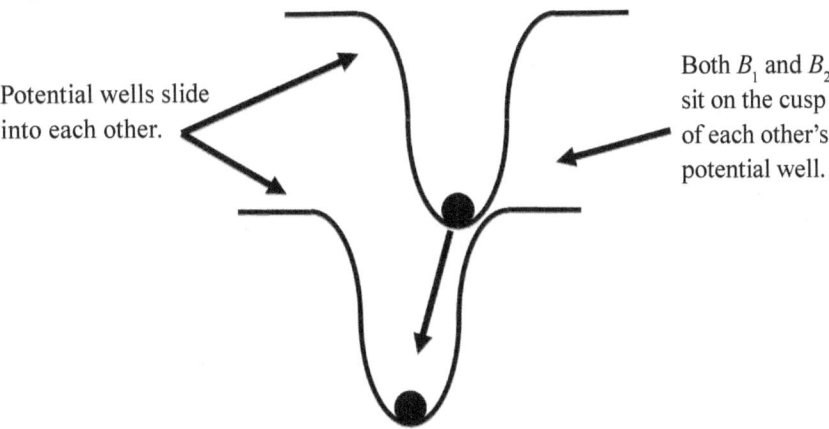

Potential wells slide into each other.

Both B_1 and B_2 sit on the cusp of each other's potential well.

● = unlike magnetic poles $B_1 \neq B_2$

The poles want to achieve a neutral state. They will both slide into each other's potential well. But in doing this, they will partly neutralise each other. A very good example of this is an unmagnetized iron bar, where for it appears not to have any magnetic properties, but when align all the domains (magnetic fields around each atom) the bar becomes a magnet. This can be very easily demonstrated if you bring together two horseshoe magnets, allowing them to come together naturally, where each pole attracts its opposite. Thus, a cancellation of polarity occurs. If you place a sheet of paper on top and then sprinkle iron filings on it, you notice very little residual magnetic activity.

DARK MATTER, UNIFIED FIELD THEORY, AND UFO'S,
ARE UNDERSTANDABLE AND ACHIEVABLE.

29

CHAPTER 6

The Seventh Dimension (Time)

The seventh dimension is linked to the previous six dimensions. If you have a moving charged object, the magnitude and direction of the magnetic field will interact with the seventh dimension, Time. In the case of photons, the magnetic field travels at right angles to the direction of motion of the wave but crosses it path as it goes. It is this crossing of paths of the magnetic field and waveform that limit its speed in the direction of the waveform. From chapter 11, it appears that if magnetic field is slowed down or the direction of rotation or is changed such that it does no longer cross the path of the photon's wave form, the particle appears to travel at phenomenal speeds. What I am now trying to propose is if we change the direction of the magnetic field form cutting across the path of the wave form such the magnetic field of particle now rotates around the direction of motion, we might have a new effect happening. This effect could be affected by the strength of both the magnetic and electric field, frequency, speed of particle, and the direction of rotation of the magnetic field, may result in allowing the particle to possibly go forward or backwards in time.

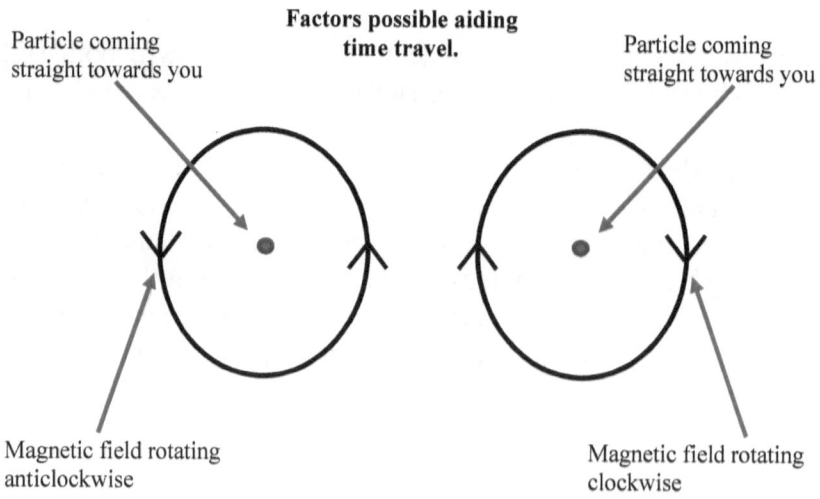

This theory could explain how some particles could be in two or more, places at the same time, and therefore be related to Entanglement theory. Therefore, when we observe entanglement events, what we are observing is one particle moving either or forwards or backwards in time.

DARK MATTER, UNIFIED FIELD THEORY, AND UFO'S,
ARE UNDERSTANDABLE AND ACHIEVABLE.

31

CHAPTER 7

Photons

Photons can be created if you allow a negatively charged electron (e^-) to collide with a positively charged electron, called a (positron) (e^+). This collision creates two photons, both of which have zero rest mass. However, before the collision both particles have masses which are measurable and finite.[11]

Arthur Ashkin and Joseph Dziedzic, scientists at Bell Labs, were the first to levitate a glass bead using an optical laser. [12] The significance of this was that the radiation pressure from the photon's momentum could be used to lift the glass bead, but it also indicated that there was far more to the photons (*Popular Science*, January 1972, p 40). This radiation pressure contributed to the concept that photons had the property of particles.

The best explanation is that as the photon passes through atomic lattice of the particle, the charges that exist within the photon interact with the surrounding electron and protons. The net effect is that the surrounding charges either attract or repel the photon, causing it to lose energy and slow down. The net effect is to in part or transfer energy to the glass particle, causing it to move and then levitate. When you stop both the electron and the positron, you can measure their mass and charge exactly. The fascinating thing about this is that this is not possible for a photon. Therefore, there is a clear distinction between the photons and electrons. Any particle that has a measurable mass is what I call 'real mass'. If you then try and accelerate real mass to the speed of light (c), you will find it will never happen. It is impossible as its mass becomes infinite.

In an attempt to explain this:

When an electron collides with a positron, they do something very strange. They first exchange half their charge in an attempt to cancel each other's charge, and then each of these half-charged pair produces a new particle called a photon.[10] The two new particles then have the following properties:

1 They appear not to have any rest mass, i.e. rest mass = 0.
2 These particles can travel at the speed of light (c).
3 When you place them against most charges, they appear not to be affected.

High energy experiments have shown that you can force two photons to re-exchange their charges and create an electron (q^-) and a positron (q^+) again. This ability to flip from a real mass to a rest mass of zero and back again indicates that the charges are not cancelled in the photon state.

When the half charges of a photon slide into each other's potential well, there is an attempt to cancel each other and become neutral. See figure 13.

Figure 13: Neutral Charge

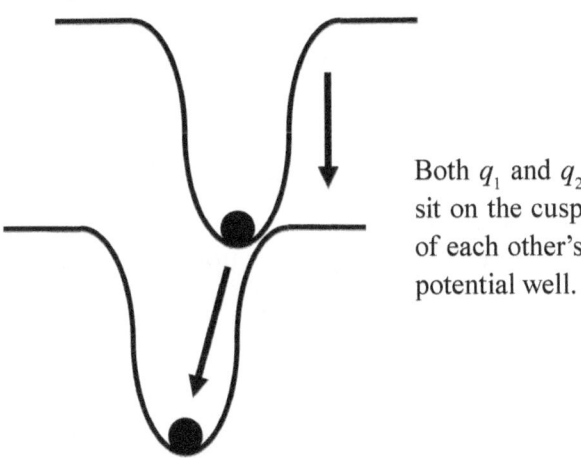

Both q_1 and q_2 sit on the cusp of each other's potential well.

DARK MATTER, UNIFIED FIELD THEORY, AND UFO'S, ARE UNDERSTANDABLE AND ACHIEVABLE.

33

But the two charges continuously try to cancel each other within the photon. This creates a situation where the electric fields are continuously trying to recreate themselves (and at the same time cancel each other out), but they are forever being pulled back on to the surface of the photon. The end result is to have two intense electric fields, acting at right angles to its surface, covering or cloaking the surface of the photon. See figure14. The net result of this effect is to blanket the photon with a fifth-dimension field effect, shifting the existence of the photon out of the fourth-dimensional space into the fifth dimension, thereby altering its existence and cutting off the influence of gravity on its mass and giving it the properties of a known photon.

Figure 14: Cloaked photon

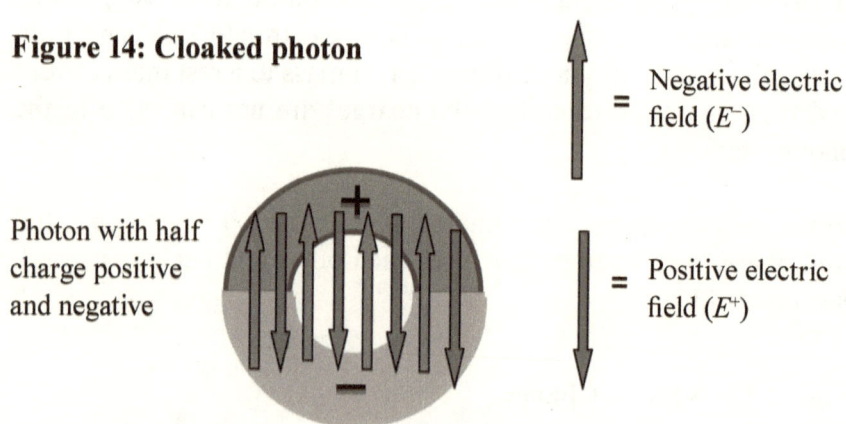

Photon with half charge positive and negative

= Negative electric field (E^-)

= Positive electric field (E^+)

The photon particle is not stationary. Due to the spinning of the positive and negative charge around the central axis results in the opposites sides of the photon travelling in opposite directions, subsequently there is a resultant addition of the magnetic field components, creating a common North and South Pole. See figure 15.

Figure 15: Common poles

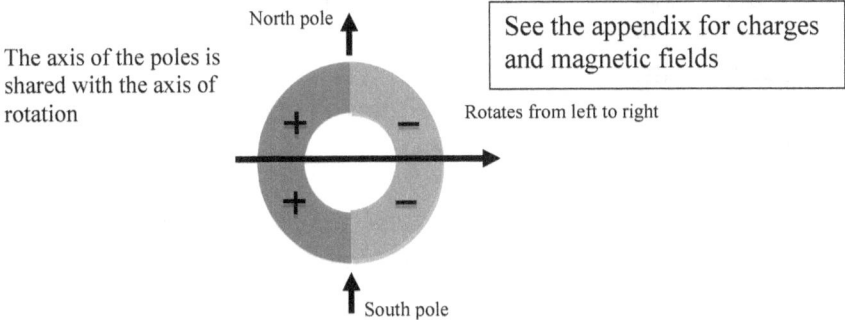

The axis of the poles is shared with the axis of rotation

North pole

See the appendix for charges and magnetic fields

Rotates from left to right

South pole

Since photons contain both a negative and a positive charge component, they will be both attracted and repelled by normal electric fields. Therefore, the net change of motion of photons would be unaffected. The two field effects acting together means that photons will not easily interact with static charges. But under the right set of circumstances, like that which occur between electrons and protons in the atom, i.e. materials such as glass, where there are a large number of protons and electrons, and their respective electrically positive (q^+) and electrically negative (q^-) fields all act at the same time on the photon.

This will lead to a dragging or slowing effect on the photons. See figure 16.

Figure 16: Dragging effect on photon

Photon flips to adjust to charge pairing in atom

An atom's nucleus with large positive charge

Attraction of nucleus

Motion of Photon

An atom's electron shells with large negative charge

Attraction of shell

Note: Electrons travel in waves in their orbits around the nucleus.

DARK MATTER, UNIFIED FIELD THEORY, AND UFO'S,
ARE UNDERSTANDABLE AND ACHIEVABLE.

35

The photon's charges are close enough to the individual atoms' respective charges and hence allow them to interact with the photon and have a dragging effect on it. The negative charge of the photon is attracted to the positive charge of the nucleus, and conversely, the positive charge component of the photon is attracted to the negative charge in the electron shells.

The photon like charges repel and unlike charges attract; thus, the positive and negative repel their like components while the unlike components attract, causing the photon to flip its orientation when passing through an atom.

Thus, the denser the medium and the more closely packed a substance is, the slower the photon will move through it. Conversely, the absence of matter has the reverse effect. This effect should see light or photons speed up when there is less matter around them or when in a near to perfect vacuum. This means that photons do not have a fixed speed and will vary, depending on the medium they are travelling in.

A possible simple experiment to demonstrate the impact of charge on photons is shown in the diagram below in figure 17. A second possible experiment is point no. 4 in the appendix.

Figure 17: An experiment to test the effect of charge on a photon:

Plates 1 and 3 are the same charge, while Plate 2 is oppositely charged.

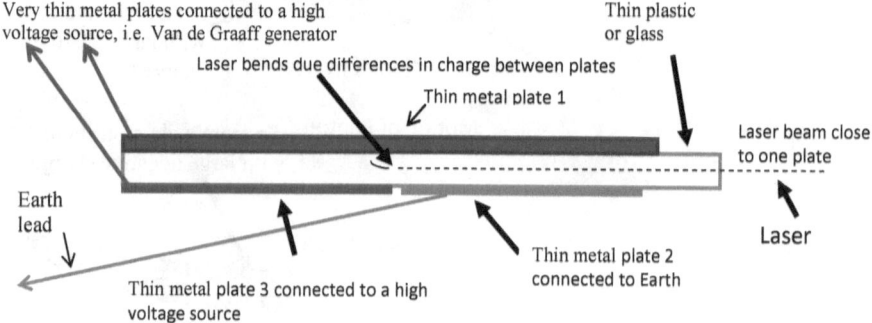

This diagram is not drawn to any scale as its intention is to show what happens to a laser light source when it passes through thin plastic or glass and is subject to uniform and non-uniform voltages. The parallel plates 1 and 2 are oppositely charged and set the orientation of the photons as they pass through the plastic or glass, but when it comes to differently charged plate 3, another like charged plate to plate 1. The photon curves towards a charged plate. The main problem, firstly, is making sure the plates have a high enough charge on them so that they will be able to interact with the photon component charges, and secondly that the insulation properties of the plastic or glass won't allow arcing between the plates. It is crucial that there is sufficient charge on each plate so that the photon can interact with the electric field. Plates 1 and 3 both have the same charge but two different voltages, but still having great enough charge to interact with the photon's charge component.

Thus, when a photon passes a large star, it is the density and intensity of the solar wind from that star which interacts with the passage of the photon, slowing it down and changing its direction. This process is very similar to what happens when light passes through a heat haze of air on the road.

Now added to this is the fact that the photon is spinning and tumbling at the same time (Dzhanibekov effect occurs due to the uneven distribution of mass within the photon). The spinning motion of the photon creates a magnetic field component which is at right angles to the motion of the wave. The tumbling motion of the magnetic field also contributes to its wavelength, due to the magnetic field interacting with the forward motion of the wave. The tumbling motion of the photon sees it flip 180 degrees and then back again. See figure 18.

Dark Matter, Unified Field Theory, and UFO's,
Are Understandable and Achievable.

37

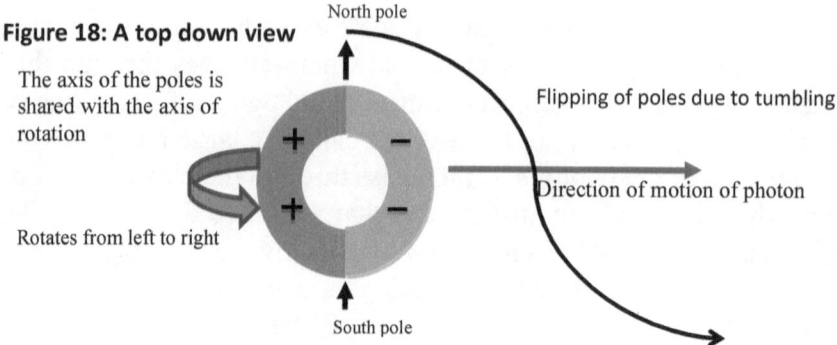

Figure 18: A top down view

North pole

The axis of the poles is shared with the axis of rotation

Flipping of poles due to tumbling

Direction of motion of photon

Rotates from left to right

South pole

Craig F. Bohren (*American Journal of Physics*, 51 (4), April, 1983, p.323–327) discovered that aluminium could absorb more photons than were radiated incident to it. He noted that, photons of 8.8 electron Volts (eV), where absorbed by aluminium and that this frequency allowed the metal to absorb more photons from the area around it than what were directly focused on it.[13]

This effect induced in the aluminium a mirroring energy signature (Resonance) which allowed for energy absorption of photons from the surrounding space which greater than the target area under investigation. This theory could also be used to create devices for power generation systems, which could absorb all photons from the surrounding area for power generation. Conversely it could reflect the photons for better energy management for power utilisation. This would have a major impact on alternative energy generation. If photon absorption becomes viable and achievable, man would have a better technique than solar cells for collecting energy. Solar cells are really not very efficient, varying anywhere from less than 6% efficiency to a maximum of 30%. The (30%) efficient solar cell is something that is found in science labs and is not commonly available.

It is also conceivable that man could create sufficiently intense negative and positive electric fields which, when pulsed correctly, would be able to 'cloak' a vehicle such as a ship. In cloaking the vehicle with these fields, he will be able to shift the vehicle from the realm of three-dimensional space into five-dimensional space. By achieving this, the rest mass of the vehicle will reduce to zero and light speed travel will be possible. Using this technology, it would be easy to travel through our world or, for that matter, our solar system. There is also the possibility of making objects invisible to the human eye and radar.

Some effects that could be observed is that higher the frequency the less voltage is required. By using a square with a very high frequency (i.e. 0.5 peta Hertz), and then slowly increasing the voltage, the object will first lose weight until it has zero rest mass. Next, depending on specific frequency, light could either be reflected or absorbed by the object. Continuing with increasing voltage will see the become transparent with very vague outline of it. The next step after that is being completely invisible and, in this state, it can travel at light speed or faster, depending on the direction of its magnetic field. It is believed for best performance that the voltage be run through a layer of conductive material just under the skin of the object. **(Note: - What is extremely important that this electric field effect must replicate that which exists in a photon.)**

Dark Matter, Unified Field Theory, and UFO's, Are Understandable and Achievable.

39

CHAPTER 8

The Tachyon and Smaller Particles

The tachyon is a particle which can travel faster than the speed of light.[14] This particle would most likely have a negative rest mass.

This particle functions in the eighth dimension. It might be possible to replicate a tachyon's field effects if we increase the positive and negative electric fields around a vehicle until they replicate those in a tachyon, while simultaneously increasing the intensity of the magnetic component and the speed of rotation around the vehicle. We must keep doing this until we produce a sympathetic field effect that is resonating at the correct frequency of a tachyon. Once it is synchronised correctly, it might be possible to have faster than light travel. If man could ever replicate the field effects around this particle, he would travel faster than light and would finally be able to move into the stars and reap the riches that may exist there. (Note; - Negative mass transit is extremely dangerous due accelerated aging.)

To summarise, there are three states of mass:

1) Real mass that can reach the speed of light   mass = 1, $\varepsilon = 1$
2) Photon's zero rest mass   mass = 0, $\varepsilon = 0$
3) Tachyon negative mass   mass = −1, $\varepsilon = -1$

One intrinsic problem we find from Einstein's and Lorentz's equations is with the state of mass of an object as it approaches the speed of light. This value cannot be equated directly to m but to the state in which m exists. In fact, the closer the velocity of the object comes to c, the more important this value becomes. To correctly adjust for this, we must add to the equations a new value incorporating the speed of the object and its state of mass. This value only becomes important when the particle's velocity approaches the speed of light.

Therefore, $v = \varepsilon\, v$, where ε is equal to 1, 0, or −1, depending on whether it is 1 for real mass, 0 for photon's rest mass, and −1 for Tachyon mass.

Applying this to Einstein's and Lorentz's equations we get:

Original equation	Altered equation

$$m = \frac{m_0}{\sqrt{(1 - v^2/c^2)}} \qquad\qquad m = \frac{m_0}{\sqrt{(1 - \varepsilon\, v^2/c^2)}}$$

$$t = \frac{t_0}{\sqrt{(1 - v^2/c^2)}} \qquad\qquad t = \frac{t_0}{\sqrt{(1 - \varepsilon\, v^2/c^2)}}$$

$$l = l_0\,\sqrt{(1 - v^2/c^2)} \qquad\qquad l = l_0\,\sqrt{(1 - \varepsilon\, v^2/c^2)}$$

Now using these equations we get the following results:

The speed of light (c) is about 299,792,458 metres per second or about 300,000 kilometres per second. Let $c = 300{,}000{,}000$ m/s.

Now let $v = .999$ of c; $m_0 = 1$; $t_0 = 1$; $l_0 = 1$; $\varepsilon = 1$. (This value is related to how the dimensions can impact an object.)

Original equation	Altered equation

$$m = 1 \div (1 - (0.999^2 \times c^2 \div c^2))^{0.5} \qquad m = 1 \div (1 - (1 \times 0.999^2 \times c^2 \div c^2))^{0.5}$$

$$\therefore m \cong 22.37 \qquad\qquad\qquad \therefore m \cong 22.37$$

DARK MATTER, UNIFIED FIELD THEORY, AND UFO'S,
ARE UNDERSTANDABLE AND ACHIEVABLE.

41

$$t = 1 \div (1 - (0.999^2 \times c^2 \div c^2))^{0.5}$$
$$\therefore t \cong 22.37$$

$$t = 1 \div (1 - (1 \times 0.999^2 \times c^2 \div c^2))^{0.5}$$
$$\therefore t \cong 22.37$$

$$l = 1 \times (1 - (0.999^2 \times c^2 \div c^2))^{0.5}$$
$$\therefore l \cong 0.0447$$

$$l = 1 \times (1 - (1 \times 0.999^2 \times c^2 \div c^2))^{0.5}$$
$$\therefore l \cong 0.0447$$

As can be seen, there is no change in velocity below the speed of light and all equations give the same results.

Now let's investigate a particle travelling at light speed.

Original equation   Altered equation

Now let $v = c$; $m_0 = 1$; $t_0 = 1$; $l_0 = 1$; $\varepsilon = 0$. (This value is related to how the dimensions can impact an object.)

$$m = 1 \div (1 - (c^2 \div c^2))^{0.5}$$
$$\therefore m = \infty$$

$$m = 1 \div (1 - (0 \times c^2 \div c^2))^{0.5}$$
$\therefore m = 1$. (This indicates mass is conserved but shifted to 5D space.)

$$t = 1 \div (1 - (c^2 \div c^2))^{0.5}$$
$$\therefore t = \infty$$

$$t = 1 \div (1 - (1 - (0 \times c^2 \div c^2))^{0.5}$$
$\therefore t = 1$. (This value shows time is unaffected and runs normally.)

$$l = 1 \times (1 - (c^2 \div c^2))^{0.5}$$
$$\therefore l = 0$$

$$l = 1 \times (1 - (0 \times c^2 \div c^2))^{0.5}$$
$\therefore l = 1$. (This value shows the length of object is unaffected.)

It can be seen that the original equations cannot cope with light speed and that they cannot explain why photons can levitate a glass bead [12].

But the altered equation does support the theory that photons do have mass. The experiment which shows photons can be used to levitate a glass bead is evidence that the photon does have momentum, which makes this experiment possible. The initial experiment on optical levitation in air was done by Ashkin and Dziedzic and has been confirmed by Susan Y. Wrbanek and Kenneth E. Weiland (NASA Glenn Research Centre, Cleveland, OH, United States, 1 January 2004, NASA/TM-2004-212889, E-14306). [12, 15]

If photons truly had zero mass, light would not be able to interact with the glass bead and it would be impossible for the glass beads to be lifted. The original equations supported zero rest mass, and this contradicts experimental evidence.

The altered equation does affect the size of the photon and therefore is unaltered, thus allowing these particles to not lose available interaction space with other particles.

Investigating Tachyons

Now let's investigate a particle travelling at light speed.

Original equation   Altered equation

Now let $v = 5c$; $m_0 = 1$; $t_0 = 1$; $l_0 = 1$; $\varepsilon = -1$. (This value is related to how the dimensions can impact an object.)

$m = 1 \div (1 - (5^2 \times c^2 \div c^2))^{0.5}$ | $m = 1 \div (1 - (-1 \times 5^2 \times c^2 \div c^2))^{0.5}$

$\therefore m = 1 \div (-24)^{0.5}$ | $\therefore m = 1 \div \sqrt{26}$

$\therefore m = 1 \div (-24)^{0.5}$ | $\therefore m \cong 0.196$

$$t = 1 \div (1 - (5^2 \times c^2 \div c^2))^{0.5}$$
$$\therefore t = 1 \div (-24)^{0.5}$$

$$t = 1 \div (1 - (-1 \times 5^2 \times c^2 \div c^2))^{0.5}$$
$$\therefore t = 1 \div \sqrt{26}$$
$$\therefore t \cong 0.196$$

$$l = 1 \times (1 - (5^2 \times c^2 \div c^2))^{0.5}$$
$$\therefore l = 1 \times (-24)^{0.5}$$

$$l = 1 \times (1 - ((-1 \times 5^2 \times c^2 \div c^2))^{0.5}$$
$$\therefore l \cong 5.099$$

From the results we can see that Einstein's and Lorentz's equations cannot cope with faster than light travel and that the results contain negative quantities which are very troublesome. Conversely, when using the altered equations, the mass relative to the earth is nearly one-fifth of the object and the time affiliated with the photon travel actually speeds up.

I find the last equation the most interesting, in that if a single point Planck particle would start vibrating at speeds greater than the speed of light, we would find the particles length would increase in direction of it motion and therefore we could create multiple dimensional strings (i.e. string theory).

If the particle is a point, we would get strings. If the particle is going in circles, it would create closed loops, the final actual shape of these strings being determined by the environment it is in. Each string has its own unique shape, which comes from the influence of its surrounding environment and the speed at which it is vibrating. This then allows a look at the building blocks of everything.

Note. Derivation of negative mass arises from Newtons Laws.

Newtons Third Law is a fundamental law in Physics and putting it in simple terms, every time a force is created there must be an equal corresponding balancing force. Or, for any force that is in action, there must be an equal but opposite one. Subsequently, for example, for a person to be able to jump they must apply a force to the ground but for their jump to be successful, the ground must supply an equal but opposite force. Taking this a step further.

$$\text{Force}(F) = \text{Mass }(m) \times \text{Acceleration }(a)$$

Subsequently, in an Initial Force (F_i) there must also a reactive Force (F_r).

∴ Force (F_i) = Force (F_r) so when you add the forces
Note, the reactive force is negative.
⇒ $F_i - F_r = 0$ (Zero)

Consequently

$$(+) F_i = (+) \{m_i \times a_i \} \quad : \quad (-) F_r = (-) \{m_r \times a_r \}$$

∴ To maintain consistency, in creating the forces, for with each positive mass their must also a negative one and extending this reasoning to positive acceleration there must also a negative one.

Finally, we come to the situation,

(+) Force	:	(-) Force
$(+) F_i = (+) \{(+)m_i \times (+)a_i \}$	:	$(-) F_r = (-) \{(-)m_r \times (-)a_r \}$
$^+F_i = (+) \{^+m_i \times {}^+a_i \}$	:	$(-) F_r = (-) \{^-m_r \times {}^-a_r \}$
$+F_i = + ({}^+m_i \times {}^+a_i)$	:	$- F_r = - ({}^-m_r \times {}^-a_r \}$

Furthering this theory, trying to find negative mass (-m) in the universe would be very difficult, because the associate negative acceleration (-a) is repulsive. Consequently, all particles would move away from every other and resulting matter becoming extremely small. These particles would exist as Plank Particles.

(**Note.** Acceleration can be proportional to velocity squared or $a \propto v^2$, subsequently apply this reasoning to negative mass we get $-a \propto -v^2$ and $-v^2$ is very important when we apply it in the Relativity Equations.)

DARK MATTER, UNIFIED FIELD THEORY, AND UFO'S,
ARE UNDERSTANDABLE AND ACHIEVABLE.

45

CHAPTER 9

Earth's Magnetic Field and Thunderstorms

How and why charges are separated in clouds:

One of the underlying principles behind why charges separate in clouds is related to Michael Faraday's work on the magnetic generator and the magnetic mono generator.[16]

Below is a diagram (figure 19) demonstrating the influence of magnetic fields on the flow of electrons. In a cathode tube, a beam of electrons travels from the cathode to the anode.

Figure 19: Magnetism on electron flow

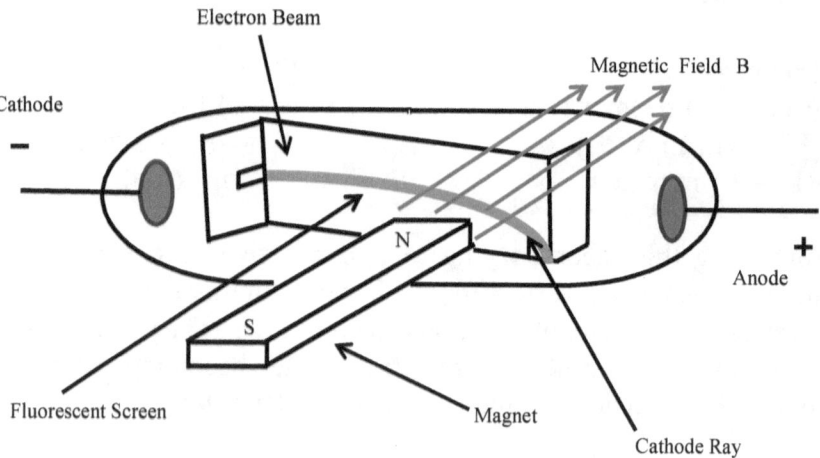

This device shows that the cathode ray's particles (electrons) move from the cathode to the anode, and as they encounter the magnetic field from the north pole of the magnet, the electrons are pushed downwards [17].

From the **oil drop experiment** by Robert A. Millikan and Harvey Fletcher in 1909, to measure the elementary electric charge of an electron, they also found that electromagnetic radiation can also create a charge in the air. [18, 19] In other words; radiation can cause the air to become ionised. Added to this is the normal frictional force in air movement, which can also ionise the air.

We can now relate the above information to the charge formation in clouds.

It is important to state here that as you move from either of the poles to the equator, you rotate with the surface of the Earth. Your surface speed at the equator will be about 465 metres per second. (Note that this is a lot faster than the speed of sound, which is approximately 332 m/s at 0°C or 344 m/s at 20°C.)

But generally, clouds move at various speeds, and their speed can be added to that of the earth's rotational speed. [20, 21]

The earth creates its own magnetic field, and in its creation, the speed of the magnetic field is equal to the speed of light. On combining this information with Michael Faraday's work on mono-pole generators,[16] we can get charge separation in our atmosphere. Drawing this relationship, we get the following diagram in figure 20. The earth's magnetic field (B) travels from the south magnetic pole (S) to the north magnetic pole (N). Now the earth rotates from the west (W) to the east (E).

DARK MATTER, UNIFIED FIELD THEORY, AND UFO's,
ARE UNDERSTANDABLE AND ACHIEVABLE.

47

Figure 20: Charge interaction in earth's magnetic field

(See the appendix for charges in magnetic a field and the left & right hand rule.)

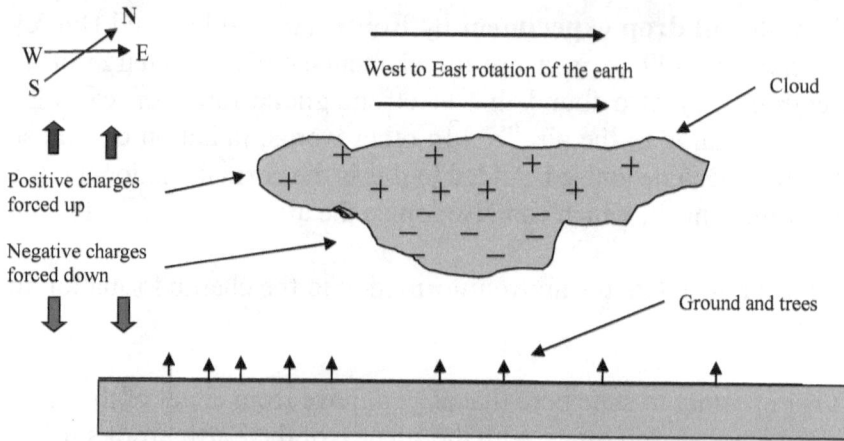

The earth's magnetic field *B* separates the charges

The net effect of this on earth's atmosphere is to separate charges, moving negative charges closer to the earth's surface and pushing positive charges away from the earth's surface. Thunderstorms store and build up a large negative charge in clouds near the earth's surface and a corresponding level of positive charge in the higher altitude clouds. This potential difference is kept by the earth's magnetic field until voltages get large enough for arcing (lightning) to occur.

This interaction of magnetic fields with the electric fields of the free moving charged particles is what causes the negative charges to be encouraged to move towards the earth and the positive charges away from the earth. (This is called the 'left-hand rule' for charges in a magnetic field.) See the appendix.

This effect can also happen in wires if the charged particle is free to move. This interaction of magnetic fields with electric fields is basic and can also be found in the working of electric generators. Therefore, at any given moment, there exists a potential difference between the earth and the different layers in the atmosphere. Depending on the particles and the charge that is on them, layers of charge can build up in the atmosphere. When the voltages build up to large enough values, lightning strikes can occur on the earth or to higher levels above the clouds.

I have the design of a simple device which can tap into this electric potential of stored charge in the atmosphere and create an environmental-friendly electrical generator. I am hoping to find the backing one day to develop it.

DARK MATTER, UNIFIED FIELD THEORY, AND UFO'S,
ARE UNDERSTANDABLE AND ACHIEVABLE.

49

CHAPTER 10

Weather

Milankovitch's theory describes the collective effects of changes in the earth's movements upon its climate. He mathematically theorised variations in the eccentricity, axial tilt, and the precession of the earth's orbit and that they determined climatic patterns on earth. The time frame for the eccentricity is about 70,000–400,000 years, the axial tilt about 41,000 years, and the precession about 26,000 years. These time frames are not set but vary in accordance with natural factors which may impact the earth's orbit around the sun.[22] None of these cycles are a hundred thousand years so in a Ice age we may experiences all of these cycles without the Ice age turning into a warm period.

The strongest factor, which I believe contributes to Ice Ages on earth, is the time frame that it takes starting from photon production within the sun and the related energy it generates to its passage to the surface of the sun and then to earth, for from the moment of nuclear fusion within the sun, which creates the photon, to its reaching the earth's surface is about a hundred thousand years (100,000 years).[23]

There is a scientific paper where it's documented that two Ice Age had anywhere between five and eight times more carbon dioxide in the atmosphere than we have now.

Some extreme changes have occurred with the onset of an Ice Age, wherein dramatic changes in weather have occurred. Milankovitch's theories happen gradually over time.

The only factor influencing our weather, which can experience rapid and extreme changes, is our sun. The sun's solar activity and winds do have a direct impact on the earth. If you calculate the kinetic energy of the solar wind and map it against world temperature, you find a very strong correlation between them. Only since 1978 has data on solar winds been collected. This lack of information makes it very difficult to accurately predict future weather. With time and greater study, we will know. But the one thing that does worry me is that the last Ice Age finished more than 12,000 years ago and each warming period lasts about 10,000 years. At present, our sun is going into a solar minimum; how do we know that this will not lead us into the next Ice Age? [23, 24]

This is something I would like to study in much greater detail. Since another the Ice Age is likely, how can we best prepare for its coming?

Some other topics I would like to cover or research in the future are the following:

1) The impact of geological entrapment of free carbon and the effect it may have on the environment. For in the early 1600's carbon dioxide levels dropped down to 270 ppm. (Note: - At 150 ppm all plants stop working, so in all reality we are in a carbon draught right now, and no-one is doing anything about it.)

   For quite some time all deserts in the world have been growing and gradually all the carbon in the world ecosystems is slowly all being locked up within the crust of the Earth. (Note: - We are in a carbon draught now so what should we be doing now.)

2) How photon entanglement can be used to transmit messages through time

3) How to create a Tachyon gate so that transportation of goods can become instantaneous

DARK MATTER, UNIFIED FIELD THEORY, AND UFO'S,
ARE UNDERSTANDABLE AND ACHIEVABLE.

51

CHAPTER 11

Epitaph Omega

The last thing I would like to finish with is the actual photon waveform. It is taught in schools, and commonly believed, that photons and electrons both travel as sine waves. In addition, this is supported by Planck's equations.[25]

Then as a photon completes a single wavelength, it also completes a circle.

On relating this circle to the wavelength:

λ = wavelength of photon which equals two multiplied by Pi multiplied by radius.

$\lambda = 2 \pi r$.

c = the speed of light, where c_c = the circumference of a circle, r = the radius, $c = \lambda f$ (the speed of a photon is equal to the wavelength multiplied by frequency), and $c_c = 2\pi r$ (the circumference of a circle is equal to 2 multiplied by Pi multiplied by radius or Pi multiplied by diameter).

On calculating the radius and rearranging the equation, note the half of the wavelength of a photon $\lambda/2$ = diameter (d) of the circle and half the waveform.

$d = 2r$ and pi = π = 3.1415926536

λ = a wavelength f = the frequency d = the diameter

$\therefore c_c = 2\pi r$ or $c = \pi 2\lambda/4$ or $c = \pi d$ given $\lambda/2 = d$.

What I would lastly like to investigate is what relationship develops when a photon loses energy and then would like to compare the speed of a photon when it is travelling in its waveform compared with its actual forward motion. See Tables 2 and 3.

All calculations were made using the Standard International System for measurements (metres, seconds, etc.).

Table 2: Speed of photon in wave compared its constant value

Values	f (Hz)	λ (m)	Speed of c	Speed in Cc
1	1.00E-44	2.998E+52	299,792,458	4.7091E+52
2	1.00E-36	2.998E+44	299,792,458	4.7091E+44
3	1.00E-28	2.998E+36	299,792,458	4.7091E+36
4	1.00E-20	2.998E+28	299,792,458	4.7091E+28
5	1E-12	2.998E+20	299,792,458	4.7091E+20
6	0.0001	2.998E+12	299,792,458	4.7091E+12
7	10000	2.998E+04	299,792,458	4.7091E+04

Because these values are so large, it becomes very difficult to see what is happening. So, to simplify, I have tabulated the logs of the values of wavelength and the log of the ratios of speed of light (*c*) against itself and that of the speed of a photon in its passage along its own wavelength. See table 2.

Table 3: Simplified values of table 2

Values	Log ë	C / C	C_c / C
1	52.5	1	6.21
2	44.5	1	5.27
3	36.5	1	4.33
4	28.5	1	3.38
5	20.5	1	2.44
6	12.5	1	1.50
7	4.5	1	0.55

By plotting these results, we get the following graph (Graph 2).

Graph 2: Ratios of *c* versus wavelength

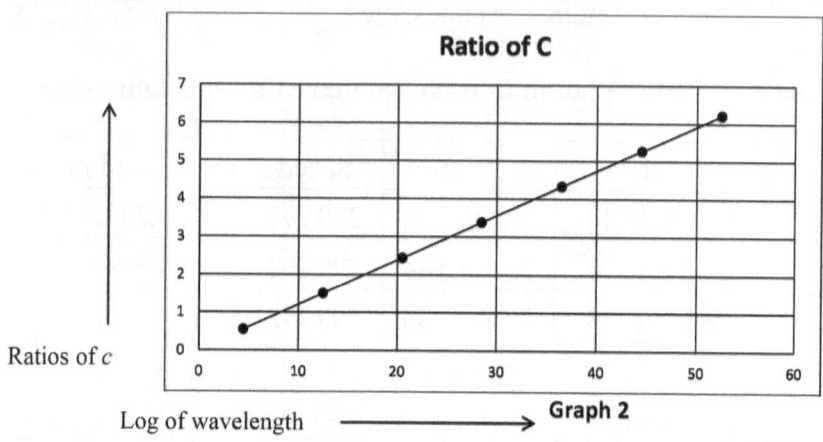

From this graph, as the photon loses energy its wavelength gets larger and in so doing so it appears to be travelling faster than light.

Conclusion:

A possible contributing factor to the loss of energy in a photon is related to the tumbling motion of the magnetic field of a photon. As previously mentioned, the Photon tumbles because of to the uneven distribution of mass within the photon (Dzhanibekov effect), so as the photon spins; this sets up **gyroscopic forces** which want to keep it going. Therefore, as a photon tumbles, it has to overcome the **gyroscopic force,** and in so doing, it loses energy, but only very slowly over a long period of time. [26]

Einstein's concept that the speed of light is a constant and does not change means that this would also impact the speed of a photon in its movement along its waveform. **This most thought-provoking**

point is that if light truly had a fixed speed, then the wavelength and frequency of light/photons would be fixed at zero, and that does not happen.

In fact, the frequency and wavelength should not be present, and the particle should be travelling as a straight line. By having different wavelengths and frequencies consideration must be given to the different speeds that could exist within the wave.

A simple look at the electromagnetic spectrum shows that photons can have nearly any wavelength or frequency, but their combined values relate to their velocity.[25]

$$c = \lambda f$$

It can be seen from these results that as a photon loses energy, its speed in the wave increases. This contradicts the laws of relativity. It appears that the first third dimensions limit themselves by their existence and act directly against the fabric of the other dimensions or the universe. Thus, for normal matter it is impossible to go faster than the speed of light unless you shift their plane of existence to higher dimensional planes.

But from Graph 2, it can be speculated that as a photon loses energy to move as a wave, it can achieve amazing speeds. It can also be noted that as the axis of rotation (Dzhanibekov effect or Tumbling) of the photon (with its **gyroscopic forces**) is slowly brought to an angle of 90 degrees to the direction of the motion of the particle, the photon appears to be able to travel at speeds that are faster than light. At the same time, the magnetic field or the magnetic moment also lines up with the axis of rotation. Thus, we have the magnetic field acting in the same direction as the tumbling motion of the photon. As the photon tumbles it reacts with other dimensions resulting in giving the photon a direction to its motion and also limits speed.

The implication of the magnetic field, with the tumbling rotation of the photon and the direction of motion of the particle all aligned (all

Dark Matter, Unified Field Theory, and UFO's,
Are Understandable and Achievable.

55

going in the same direction), means photons speed will be affected. But if we change the direction of the tumbling and make it rotate around the photon, while making sure it does not cut the path or the direction of movement, it can then interact with time. As stated earlier, to access the next dimension, it must be at 90 degrees to all the previous dimensions and all lower dimensions must be involved. Both the north and south field are aligned with the normal rotational axis of the Photon. Thus, under the right conditions, or how the north and south magnetic fields interact with the seventh dimension, time, particles will be seen moving forwards or backwards in time. Therefore, it is possible for one particle to be in two different, places at the same time, but in fact it is at two different points in its own timeline (Quantum entanglement).

Quantum Entanglement

Coming back to the wavelength, if the tumbling motion of a photon is not aligned with, or removed from the photons motion, then it would be able to travel at greater than the speed of light (See graph 2). Therefore, we might have two types of tachyons. If so, the two types of tachyons might be, –

1. One with a rest mass of zero or $m = 0$
2. The other with negative value of mass or $m = -1$.

Special note, for above to occur, both the magnetic field component and that the axis of rotation of charges is are at right angles to the observable motion of the photons, wave.

All man, has to do, is create a device that can mimic these conditions and we can build machines that travel at light speed or faster or maybe become invisible or travel through time. These devices would allow us to build generators that could absorb energy from the universe itself. Creating (Tachyon?) gates with linked resonance and teleportation of real matter might be possible.

To finish with, I will briefly discuss photon/electron interactions in electron orbits. Since an electron is negatively charged and a **photon** contains both a positive and a negative charge, when they come into close, proximity of each other, the opposite charges will attract. The end, result is partial shielding of the nucleus. The greater the energy, the greater the shielding, which allows the electron to move away and have a greater distance away from the nucleus. This shifts the electron to new orbits or ionise the atom completely. Whenever an electron moves to a closer orbit, there is a corresponding release of photons to the specific energy component. By forcing the electron in a Hydrogen to closer to the proton you start to see electric field oscillations similar to that which exist a photon and once this any number of things can happen, for example the absorption of photons (energy) from elsewhere, repulsion of light or other Photon, behave like dark matter or a proton, the proton with the electron starting to lose their mass.

These effects could lead to the possible conclusion that stationary or near stationary photons could be teleported at phenomenal speeds. Thus, as from Craig F. Bohren (*American Journal of Physics*), in the creation of appropriate electrostatic frequencies photons can be absorbed from elsewhere. [13]

This effect means also that certain particles can be transmitted and reabsorbed and transmitted again and again and again. But·they could be all the one particle but at different points in its own timeline, such that if anything happens to the particle, it can be observed in the future or the past or the present. Note that distance is not a key factor in this effect.

Finishing

The Ideas I have developed are my own and I have developed then over 38 Years.

There is ample proof for the theory in my book, and I shall cover some of them now.

1) When two black holes collide a portion of their mass will vanish. For example, one black hole is 30 solar masses and the other would be 10 solar masses, the resultant back hole should be 40 solar mass but what happens is a black hole **of 37 solar masses is created.** No, the energy would not be utilized in a gravitational wave, but the solar mass would be converted into the production of **Negative Mass** ("The build-up of negative mass in the black hole is what causes the black hole to instantly expand/explode?").
(Note: - Accelerating lead to high speeds and then crashing it into itself does not create new gravitational waves but a gravitational wave correction of the sum of the new mass being created.)

2) If you investigate black holes you should fine intense electrostatic and magnetic fields at and around the equator of the Black Hole.

3) If energy does escape from a black hole it would as high energy photons and it would be at the poles of a Black Hole where are there are weaker electrostatic forces.

4) If you look at the structure of the universe it should look like Swiss cheese. So, when you take a section or a slice of swiss cheese, the cheese section of the cross-section would be all the Galaxies in the universe while the holes are dark and could be the presence of either black holes or the former existence of black holes.

5) When you look deep into the universe you should able to observe in the universe isolate stars or remnant stars that are older than the initial big bang.

6) If you look at white light over a very large distance, you should see a longitudinal separation of colours/ wavelengths. The shorter the wavelength the less distance the photon travels in its waveform and as compared to its equilibrium point, over time. However, as the wavelength gets larger, the further the photon has to travel in its own, wavelength. Therefore, there is a greater disparity distance covered in the wave as compared to the distance covered via the equilibrium point. Therefore, as the photon's wavelength gets larger, so to

its overall distance of travel in its waveform. Conversely this does not happen to the equilibrium point. Therefore, with a loss of energy over time (Red shift of the Hydrogen spectra), photons with wavelengths close to the equilibrium point will have a greater linear distance, conversely the reverse happens for increasing wavelengths and staged separation of waveforms will occur with time.

7) The greater the intensity/density of electrostatic charges in a medium the slower a photon will go. The presence of the Ion wind around a star is like very charged medium and which will change the path of photons, due to charge interactions between the particles. The greater the density of a substance with its associated charges (protons and electrons), the more a photon will be slowed down or refracted or reflected.

8) Photons contain positive and negative charges which are rotating in an extremely small area of space. Now if you bring together a Proton and Electron close to each other, and then spin them around each other, they will behave like a photon or dark matter (See Mike Power sun cells). The Theory previously disgust allows man the opportunity to replicate these field effects and apply them to produce vehicles that can travel at the speed of light or faster. (Note, Higher the frequency used the lower the voltages required to make vehicle work.)

9) Dark Matter = Negative mass (10^{-38} m) with negative gravity, Photons and matter that copies the field effects of photons, matter containing both real mass and negative mass particles. (Note; - The interaction between real mass and negative mass means that galaxies would contain twice the amount of Negative mass particles than the space which exist between galaxies and therefore with enough distance galaxies would stop attracting each other and start repelling each other).

10) To finish my book also includes the existence of strings from string theory and how the earth itself is a massive generator that can be tapped into.

Most of the subject matter I have listed as potential proof of my theories has already been discovered and verified by other scientists.

In future work, I will endeavour to relate the weak and strong nuclear forces to Planck's particles and electrostatic charge. I will also discuss further tachyons and other atomic and subatomic interactions.

APPENDIX

1 Charges and their interaction with a magnetic field

To show that negative and positive charges can produce similar fields, see the diagram below (Figure 21). The arrows indicate the direction of motion of particles. When we have q^- and q^+ entering a magnetic field from opposite directions, we get the below results.

Figure 21: Interaction of charged particles with a magnetic field:

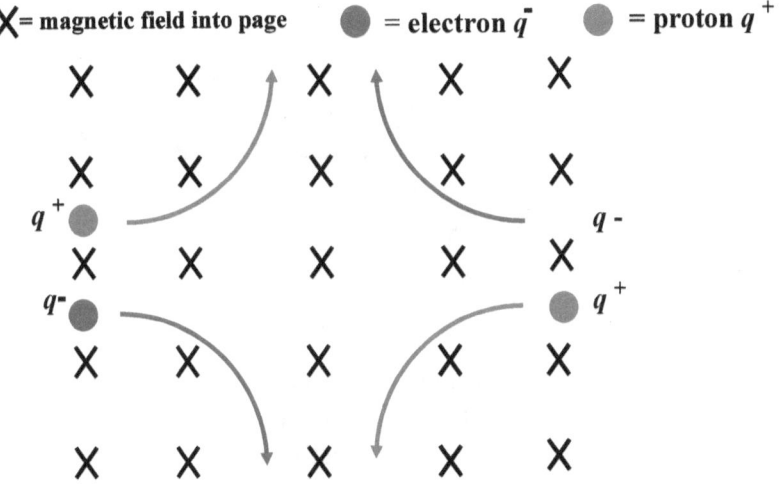

Using the left-hand rule for the negative and the right-hand rule for positive charges, we get both particles coming from opposite directions, but interacting with the magnetic field in the same way. Therefore, they leave the magnetic field in the same direction.

But when the charges enter from the same direction, they leave in opposite directions. (See points 6 and 7 for left- and right-hand rules.)

2   When a positron and electron combine their charge, potentials move into each other's potential well, and they produce two new particles which is about 1,000,000 smaller than the diameter of an atom or about 1.37×10^{-16} m.

3   A hydrogen atom has a diameter of about one Armstrong = 10^{-10} m.

4   A second possible simple experiment to demonstrate the impact of charge on photons is given in the diagram below (Figure 23).

Figure 22: Second experiment testing charge interaction with a photon

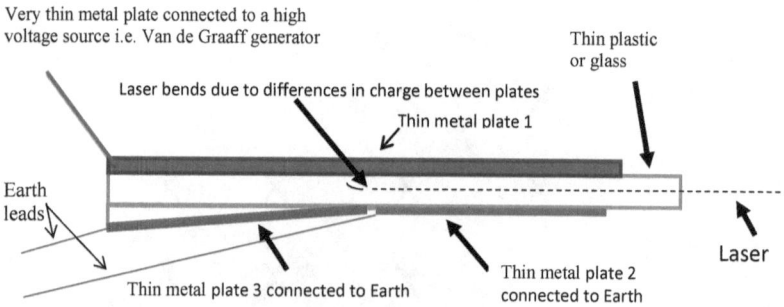

This diagram is not drawn to any scale as its intention is to show what happens to a laser light source when it passes through thin plastic or glass and is subject to uniform and non-uniform voltages/charges. The initial parallel plates set the orientation of the photons as they pass through the plastic or glass, but when it comes to plate 3, an angled lesser charged plate, a photon curves towards the greater charged plate. The main problem firstly is to make sure the plates have a high enough charge on them so that they are able to interact with the photon component charges and secondly that insulation properties of the plastic or glass won't allow arcing between the plates. It is crucial that there is sufficient charge on each plate so that the photon can interact with the electric field. The angled plate must have a lesser charge on it but still great enough charge on it to still interact with the photon's charge component.

$$\lambda = \frac{h}{mc} = \frac{2Gm}{c^2} \qquad \text{Planck's equation.}$$

5   Given that tachyons must have a negative mass (m^-), it can be extrapolated that there must be a corresponding gravitational field to go with it or gravity plus (g^+) or repulsive. **Normal gravity or gravity minus (g^-) shows that all objects with mass are attracted to each other, but with gravity plus (g^+) all matter is pushed away from each other. (Evidence for repulsive gravity may be found in why galaxies are moving away from each other at every increasing speed.)** Note that both these fields affect dimensions in exactly the same way, but they mirror or are opposite to each other.

The gravitational field behaves in a similar way to electric fields, for example, the negative electric field (q^-) or positive electric field (q^+) and how they alter a dimension. But a fundamental difference is in how the charges interact with each other, **for same charged particles repel and opposite charges attract.** On comparing a gravitational field with an electrostatic field, we find that they behave in an opposite way to each other.

To create an equation to relate these fields, we must look at how they affect the dimensional space. The first three dimensions remain the same, but for the fourth dimension we find with g^+ and g^-, are **laterally inverted**. Now the inverse relationship does not truly describe them. Now to summarise this information and put it into an equation, see below.

Let $\theta = 90°$;   let G^+ = modulus of gravity.

                                 G^- = modulus of gravity; $\therefore G^+ = G^-$

                                   q^- = modulus of a negative charge

                                   q^+ = modulus of a positive charge

DARK MATTER, UNIFIED FIELD THEORY, AND UFO'S,
ARE UNDERSTANDABLE AND ACHIEVABLE.

63

$$G^- \times \frac{1}{G+} \times \sin \theta = 1$$

$$q^- \times \frac{1}{q+} \times \sin \theta = 1$$

$$G^- \times \frac{1}{G+} \times \sin \theta = q^- \times \frac{1}{q+} \times \sin \theta.$$

To relate gravity and electric field effects with each other, you must always take into account that both these fields do have a positive and negative aspect and are laterally inverted. The influence or counter mirroring fields both act in the same way. Thus, it is simply not just one quantity by itself. As with Coulomb's Law, all factors which may be essential to the equation must be included if you are trying to calculate relationships.

6   The left-hand rule is for used for determining the force on a negative charge in a magnetic field. But it will show you which direction a negative charge will take in a magnetic field.

Direction of charge movement
B = Direction of magnetic field
F = Direction of force on to the charge.

The thumb points in direction of negative charge movement

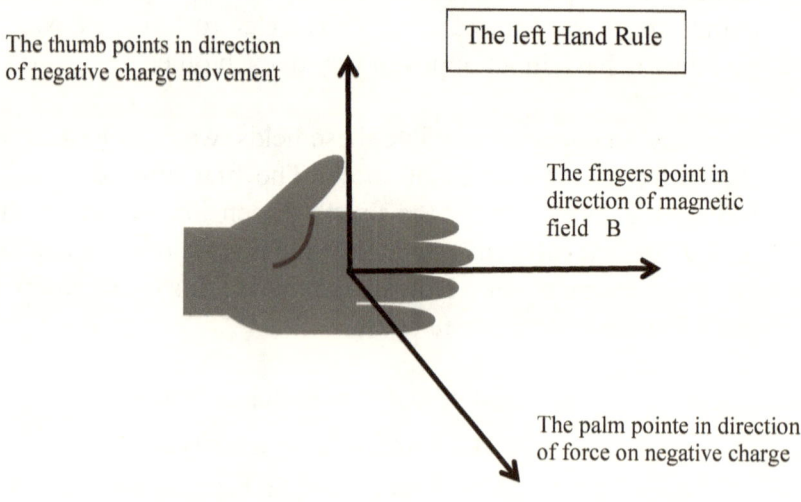

The left Hand Rule

The fingers point in direction of magnetic field B

The palm pointe in direction of force on negative charge

7   The right-hand rule is used for determining the force on a positive charge in a magnetic field. But it will show you which direction a positive charge will take in a magnetic field.

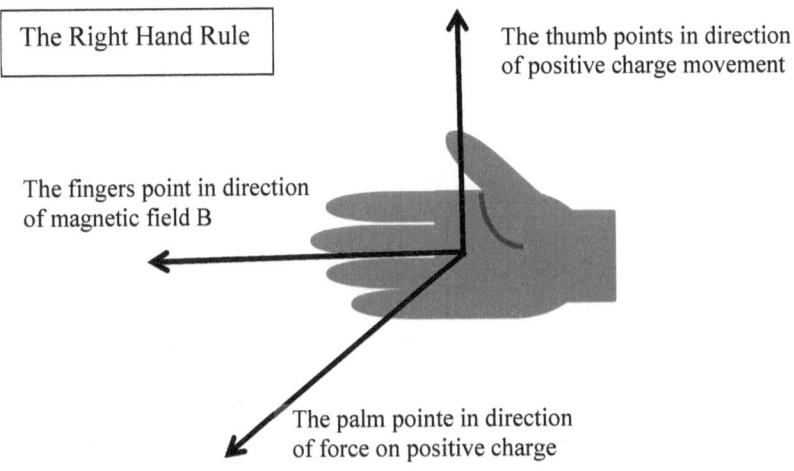

| The Right Hand Rule |

The thumb points in direction of positive charge movement

The fingers point in direction of magnetic field B

The palm pointe in direction of force on positive charge

8   The size of an electron

Taking the Diameter of the proton to be 1.75×10^{-15} meters, we can calculate its volume to be using the equation $V = 4/3 * \pi * r^3$, and therefore the Volume equals $2.81 \times 10^{-45} \text{ m}^3$. The approximate mass of an electron is about 1836.15267343 smaller than a proton. Solving for the volume of an electron we get $1.53 \times 10^{-48} \text{ m}^3$. Resolving for the volume equation for $r = (3v/4 \pi)^{1/3}$, we get that elections radius being equal to 7.14646×10^{-17} meters and hence the diameter $(d = 2 \times r)$ is 1.42929×10^{-16} meters.

9   **Plank particles are about 10^{-38} of a Metre.**

From Newton's 3rd law, for every action there is an equal but opposite one.

Therefore, when a force is created, there should also be a mirror or a Reactive opposite force.

DARK MATTER, UNIFIED FIELD THEORY, AND UFO'S, ARE UNDERSTANDABLE AND ACHIEVABLE.

65

Note; - As soon as you try to create a single force you are also creating its counterpart the reactive force.

And   Force (F) = Mass (m) X Acceleration (a)

Force {Initial} = Force {Reactive};
- (F_I) = (F_R) (Initial = I = 1 & Reactive = R = 2)

Newtons second Law States that a Force (F) is equal to the mass of an object (m), multiplied by its acceleration (a).

Given that $\quad$ **$F_1 - F_2 = 0$** $\quad ; \quad \therefore \quad$ **$F_1 = F_2$**

Thereby $\quad\quad$ **$F_1 = m_1 \times a_1$** $\quad ; \quad$ **$F_2 = m_2 \times a_2$**

Gravity (g) is a force that is linked to all objects that a mass. Linking gravity to force.
F_1 & F_2 $\quad\quad\quad\quad\quad$ **$F_1 = m_1 \times g_1$;** $\quad\quad$ **$F_2 = m_2 \times g_2$**
$\quad\quad\quad\quad\quad\quad\quad\quad\quad$ **$m_1 \times g_1 = m_2 \times g_2$**

But when you add the two forces they equate two Zero (0), therefore one of the two forces must be negative.

$\quad\quad$ **$(F_1) + (F_2) = 0$** $\quad\quad$ **and** $\quad\quad$ **$(F_{2)} = (-1 \times F)$**

Subsequent expansion of theory results in all the components in the equation become negative. $\quad$ **Therefore $(-1 \times F) = (-1) \times (-m) \times (-a)$**

There is a problem with this equation in that with normal mass it attracts all other matter that is like itself. Thus, over time objects become more and more massive, increasing in gravity and slowly turn into super dense black holes.

Conversely with negative mass the gravity is also negative and therefore repulsive and subsequently negative mas will always try to remain the size of a plank particle. (Note, - Making it extremely difficult to detect.) As discussed earlier negative mass can travel at high velocities, suffer quick aging and never forming into massive objects, let alone super dense ones. Because of the properties of Negative mass, they can only truly exist as plank sized particles. Subsequently, negative mass can only be created in black holes and is easily observed when two black holes collide, and a proportion of their mass seems to vanish. The disappearance of mass is what is convert into negative mass. All black holes continue to build up negative mass to where 50 to 67 % of the black hole is negative mass. When the right conditions are generated, the black hole will then expand faster than light. About 20 to 25 % will convert back to normal matter spread out throughout the universe. To find proof of this, one should be able to observe that the structure of the Cosmos should look like swiss cheese. This is because areas will show up as black bubbles, while the cheesy part would consist of billions and billions of Galaxies.

DARK MATTER, UNIFIED FIELD THEORY, AND UFO'S,
ARE UNDERSTANDABLE AND ACHIEVABLE.

67

REFERENCES

(Note all the readings were re-explored in August 2014.)

1. '$E = mc^2$: Deriving the equation',
   http://www.emc2-explained.info/Emc2/Derive.htm#.U9Sd
   dXkcSUk.

2. Craig F. Bohren Experiment, How a particle more than the light
   incident on it, *American Journal of Physics*, April 1983, vol. 51,
   323–326.

3. Bohren, C. F. (1983). How can a particle absorb more than the
   light incident on it? *American Journal of Physics, 51*(4), 323-327.
   doi:10.1119/1.13262

4. Shuntaro Takeda, et al., Deterministic quantum teleportation
   of photonic quantum bits by a hybrid technique (Quantum
   Teleportation over 143 kilometres Using Active Feed-forward),
   Nature 500, 15 August 2013, p. 315–318. doi: 10.1038/nature12366
   published online 14 August 2013.

5. Red shift (Doppler effect), Electric charge, http://en.wikipedia.
   org/wiki/Electric_charge.

6. Dimension (mathematics and physics),http://en.wikipedia.org/
   wiki/Dimension_(mathematics_and_physics)

7. Gravity, http://en.wikipedia.org/wiki/Gravitation

8. Electric charge, http://en.wikipedia.org/wiki/Electric_charge

9. Electric field, http://en.wikipedia.org/wiki/Electric_field.

10. Coulomb, http://en.wikipedia.org/wiki/Coulomb.

11. Magnetic field, http://en.wikipedia.org/wiki/Magnetic_field.

12. Pair production (q^+ & q^-) from gamma ray, http://www.britannica.com/EBchecked/topic/438692/pair-production.

13. Ashkin, A., 'Acceleration and trapping of particles by radiation pressure', *Phys. Rev. Lett.* 24 (4), 1970, 156–159.

14. Craig F Bohren Experiment, How a particle more than the light incident on it (American Journal of Physics, April 1983), vol. 51, p 323-326

15. Tachyon faster than light, http://en.wikipedia.org/wiki/Tachyon.

16. Ashkin A., Dziedzic J. M., Bjorkholm J. E., Chu S. 'Observation of a single-beam gradient force optical trap for dielectric particles', *Opt. Lett.* 11 (5), 1986, 288–290.

17. Michael Faraday, http://en.wikipedia.org/wiki/Michael_Faraday

18. Cathode ray tube, http://en.wikipedia.org/wiki/Cathode_ray_tube.

19. Millikan's oil drop experiment, http://amrita.vlab.co.in/?sub=1&brch=195&sim=357&cnt=1.

20. Millikan's oil drop experiment, http://en.wikipedia.org/wiki/Robert_Andrews_Millikan.

21. Earth, http://en.wikipedia.org/wiki/Dimension_(mathematics_and_physics).

22. Earth's rotational speed, http://en.wikipedia.org/wiki/Earth's_rotation.

23. Milankovitch cycles, http://en.wikipedia.org/wiki/Milankovitch_cycles.

24. The sun: The average time 100,000 years for photons production; http://qi.com/infocloud/the-sun.

25. Zbigniew Jaworowski, M.D., Ph.D., D.Sc., 'Ice coming not global warming, Solar Cycles, Not CO2, Determine Climate'; http://www.google.com.au/url?sa=t&rct=j&q=&esrc=s&source=web&cd=1&cad=rja&uact=8&ved=0CB0QFjAA&url=http%3A%2F%2Fwww.21stcenturysciencetech.com%2FArticles%252002004%2FWinter2003-4%2Fglobal_warming.pdf&ei=ulKVVIL-DuKsmAXmpIH4DQ&usg=AFQjCNE5D_wrO0zbekUlYU2i5Wrk9i-WmQ&sig2=fHczdYU1QWRu__4QiH3S8Q&bvm=bv.82001339,d.dGY

26. Planck's Laws $c = \lambda f$, http://en.wikipedia.org/wiki/Planck's_law.

27. Gyroscope, http://en.wikipedia.org/wiki/Gyroscope.

BIBLIOGRAPHY

(Note, all these readings were re-explored in August 2014.)

I used Wikipedia for its currency and ease of access, to refresh my past readings and knowledge. Wikipedia is known for occasional corruption of content, but all information has been previously resourced from past studies.

1. Clouds and charge, http://oceanservice.noaa.gov/education/yos/resource/JetStream/lightning/lightning.htm.

2. Dark matter, http://en.wikipedia.org/wiki/Dark_matter.

3. Einstein and Lorentz equations, http://en.wikipedia.org/wiki/Lorentz_transformation.

4. Electron–positron annihilation, http://en.wikipedia.org/wiki/Electron%E2%80%93positron_annihilation.

5. Electron, http://en.wikipedia.org/wiki/Electron.

6. Electrons travel in waves around the nucleus, http://hyperphysics.phy-astr.gsu.edu/hbase/ewav.html.

7. Elementary particle, http://en.wikipedia.org/wiki/Elementary_particle.

8. Gamma-ray photon produces an electron-positron pair, http://en.wikipedia.org/wiki/Pair_production.

9. Gravity well, http://en.wikipedia.org/wiki/Gravity_well.

10. Hall effect, http://en.wikipedia.org/wiki/Hall_effect.

11. Homopolar generator Michael generator, http://en.wikipedia.org/wiki/Homopolar_generator

12. Homopolar generator Michael generator, http://en.wikipedia.org/wiki/Homopolar_motor.

13. Lee Clippard,; Dr Mark Raizen, http://www.utexas.edu/news/2010/05/20/brownian_particles_research

14. Left-hand rule, http://www.upsbatterycenter.com/blog/wp-content/uploads/2014/07/Left-Hand-Rule.jpg.

15. Neutrino's, http://en.wikipedia.org/wiki/Neutrino.

16. Neutron, http://en.wikipedia.org/wiki/Neutron.

17. Optical tweezers and photon levitation, http://en.wikipedia.org/wiki/Optical_tweezers.

18. Photon, http://en.wikipedia.org/wiki/Photon.

19. Photon entanglement, http://en.wikipedia.org/wiki/Photon_entanglement.

20. Right hand rule for positive charges, http://upload.wikimedia.org/wikipedia/commons/6/6f/Right_Hand_Rule_vBF2.PNG.

21. Rotation around a fixed axis, http://en.wikipedia.org/wiki/Rotation_around_a_fixed_axis.

22. Solar wind, http://en.wikipedia.org/wiki/Solar_wind.

23. Sunspots, solar winds, carbon dioxide and ice ages, http://www.climate4you.com/GreenhouseGasses.htm

24. Sunspots, solar winds, carbon dioxide and ice ages, http://www.climate4you.com/Sun.htm

25. Sunspots, solar winds, carbon dioxide and ice ages, http://www.climate4you.com/GlobalTemperatures.htm

26. The electromagnetic spectrum, http://en.wikipedia.org/wiki/Electromagnetic_spectrum.

27. Theory of relativity, http://en.wikipedia.org/wiki/Theory_of_relativity.

28. Zbigniew Jaworowski Iceage not global warming; http://www.warwickhughes.com/icecore

INDEX

L

laser 36
laser beam 13, 36
left-hand rule 47, 61
Lorentz's equations 9, 40-1, 44

M

magnet 45-6
magnetic component 26, 40
magnetic field 26-7, 35, 37-8, 45-7,
        53-4, 61, 64-5, 70
  components 34-5, 37-8, 55
  earth's 45-7
  interaction of 47
magnetic generator 45
magnetic monopole 26
magnetic source 27-8
mass 9-11, 17-20, 22-3, 32, 34, 40,
        42-4, 62
  definite 17
  discrete 18
  measurable 32
  negative 20, 40, 62
  point 18-19
  real 32-3, 40-1
  tachyon 41
  zero 43
mass dilation 9
mass dilation equation 10
matter 12-13, 17, 36, 39, 62, 73
  dark 73
  normal 54
  ordinary 10
  real 55
Michael Faraday's work 45-7
Milankovitch's theory 49

Millikan, Robert A. 45-6
modulus 17, 20, 63
motion 11, 26, 35, 37-8, 54-5, 61
  forward 37-8, 52
  spinning 37-8
  tumbling 37, 53, 55
multiple dimensional strings 44

N

negative charge 23, 35-6, 47, 63-4
negative rest mass 40
net effect 13, 47
neutral state 25, 29
neutrinos 74
neutrons 17, 19, 74
nuclear forces, strong 59
nuclear fusion 49
nucleus 35-6, 56, 73

O

optical levitation 43

P

particles 9, 12, 17, 20, 32-4, 40, 43-4,
        48, 53-6, 61, 69-70
  cathode ray's 45-6
  new 33, 61-2
  plank 22
  single point Planck 44
  subatomic 17, 19
photon waveform 51
photons 9-13, 32-8, 40, 42-4, 49, 51-
        6, 62, 74
  absorption 38
  charge component 36-7, 62

DARK MATTER, UNIFIED FIELD THEORY, AND UFO'S,
ARE UNDERSTANDABLE AND ACHIEVABLE.

79